· PERFE

Pineapple

RECIPES

TROPICAL FAVORITES, SIDES, SWEETS, AND SIPPERS

pil

Publications International, Ltd.

Some of the products listed in this publication may be in limited distribution.

Pictured on the front cover: Easy Upside Down Cake (*page 60*).
Pictured on the back cover (*left to right*): Marinated Pineapple Dessert (*page 52*) and Tropical Chicken Wings (*page 15*).

ISBN: 978-1-68022-620-1

Manufactured in China.

8 7 6 5 4 3 2 1

Microwave Cooking: Microwave ovens vary in wattage. Use the cooking times as guidelines and check for doneness before adding more time.

CONTENTS

SIMPLE SIDES

Fruit Salad
with Creamy Banana Dressing
—————— Makes 8 servings ——————

2 cups fresh pineapple chunks

1 cup cantaloupe cubes

1 cup honeydew melon cubes

1 cup blackberries

1 cup sliced strawberries

1 cup seedless red grapes

1 medium apple, diced

2 medium ripe bananas, sliced

½ cup vanilla nonfat Greek yogurt

2 tablespoons honey

1 tablespoon lemon juice

¼ teaspoon ground nutmeg

1. Combine pineapple, cantaloupe, honeydew, blackberries, strawberries, grapes and apple in large bowl; gently mix.

2. Combine bananas, yogurt, honey, lemon juice and nutmeg in blender or food processor; blend until smooth.

3. Pour dressing over fruit mixture; gently toss to coat evenly. Serve immediately.

Gingerbread Pineapple Muffins

Makes 24 servings

1 can (8 ounces) crushed pineapple in juice, undrained

1 package (14.5 ounces) gingerbread cake and cookie mix

¾ cup lukewarm water

1 egg

2 teaspoons canola oil

¼ cup chopped walnuts (optional)

Powdered sugar (optional)

1. Preheat oven to 350°F. Spray mini muffin pan with nonstick cooking spray.

2. Place pineapple with juice in fine sieve over medium bowl; drain well, pressing pineapple to release juices.

3. Combine gingerbread mix, water, pineapple juice, egg and oil in large bowl; whisk 2 full minutes, scraping side often. Stir in walnuts, if desired.

4. Spoon batter into prepared muffin cups. Top each with equal amount of pineapple. Bake 13 to 16 minutes or until toothpick inserted into centers comes out clean.

5. Remove to cooling rack; cool completely. Sprinkle lightly with powdered sugar, if desired.

Note: After these muffins are completely cooled, they can be stored in an airtight container for up to 2 days.

Pineapple Ham Fried Rice

—— Makes 4 to 6 servings ——

8 ounces cooked smoked ham steak

3 tablespoons vegetable oil, divided

2 tablespoons sliced almonds

1 small green bell pepper, cut into strips

2 green onions, coarsely chopped

4 cups cooked rice, cooled

1 can (8 ounces) pineapple chunks packed in juice, undrained

2 tablespoons raisins

2 to 3 tablespoons reduced-sodium soy sauce

1 tablespoon dark sesame oil

1. Cut ham into 2-inch strips; set aside.

2. Heat wok over medium-high heat 1 minute or until hot. Drizzle 1 tablespoon vegetable oil into wok and heat 30 seconds. Add almonds; stir-fry until golden brown. Remove from wok.

3. Add remaining 2 tablespoons vegetable oil to wok and heat 30 seconds. Add ham, bell pepper and green onions; stir-fry 2 minutes. Add rice, pineapple with juice and raisins; stir-fry until heated through.

4. Stir in soy sauce and sesame oil; stir-fry until well mixed. Transfer to serving bowl. Sprinkle with almonds just before serving.

Piña Colada Muffins
—— Makes 18 muffins ——

2 cups all-purpose flour

¾ cup sugar

½ cup flaked coconut

2 teaspoons baking powder

½ teaspoon baking soda

½ teaspoon salt

2 eggs

1 cup sour cream

1 can (8 ounces) crushed pineapple in juice, undrained

¼ cup (½ stick) butter, melted

⅛ teaspoon coconut extract

Additional flaked coconut (optional)

1. Preheat oven to 400°F. Spray 18 standard (2½-inch) muffin cups with nonstick cooking spray or line with paper baking cups.

2. Combine flour, sugar, ½ cup coconut, baking powder, baking soda and salt in large bowl; mix well.

3. Beat eggs in medium bowl with electric mixer at medium speed 1 to 2 minutes or until frothy. Beat in sour cream, pineapple with juice, butter and coconut extract. Stir into flour mixture just until combined. Spoon batter into prepared muffin cups, filling three-fourths full.

4. Bake 15 to 20 minutes or until toothpick inserted into centers comes out clean. If desired, sprinkle tops of muffins with additional coconut after first 10 minutes. Cool in pans 2 minutes. Remove to wire racks; cool completely.

Gingered Pineapple and Cranberries

— Makes 4½ cups —

2 cans (20 ounces each) pineapple chunks in juice, undrained

1 cup dried sweetened cranberries

½ cup packed brown sugar

1 teaspoon curry powder, divided

1 teaspoon grated fresh ginger, divided

¼ teaspoon red pepper flakes

2 tablespoons water

1 tablespoon cornstarch

Slow Cooker Directions

1. Place pineapple with juice, cranberries, brown sugar, ½ teaspoon curry powder, ½ teaspoon ginger and pepper flakes in 1½-quart slow cooker. Cover; cook on HIGH 3 hours.

2. Combine water, cornstarch, remaining ½ teaspoon curry powder and ½ teaspoon ginger in small bowl; stir until cornstarch is dissolved. Add to pineapple mixture. Cook, uncovered, on HIGH 15 minutes or until thickened.

Variation: Substitute 2 cans (20 ounces each) pineapple tidbits in heavy syrup for pineapple and brown sugar.

TROPICAL FAVORITES

Tropical Chicken Wings
—— Makes 6 to 8 servings ——

1 jar (12 ounces) pineapple preserves

½ cup chopped green onions

½ cup soy sauce

3 tablespoons lime juice

2 tablespoons pomegranate molasses *or* honey

1 tablespoon minced garlic

2 teaspoons sriracha sauce*

¼ teaspoon ground allspice

3 pounds chicken wings, tips removed and split at joints

1 tablespoon toasted sesame seeds

Sriracha is a spicy sauce made from dried chiles that is used as a condiment in several Asian cuisines. It can be found in the ethnic section of major supermarkets, but an equal amount of hot pepper sauce may be substituted.

Slow Cooker Directions

1. Combine preserves, green onions, soy sauce, lime juice, pomegranate molasses, garlic, sriracha sauce and allspice in slow cooker; stir well.

2. Add chicken wings; stir to coat. Cover; cook on LOW 3 to 4 hours or until wings are fork-tender.

3. Sprinkle with sesame seeds just before serving.

Tip: Pomegranate molasses is a syrup made from pomegranate juice cooked with sugar. You can easily make your own if it isn't in the ethnic foods aisle of your local supermarket. For this recipe, bring ½ cup pomegranate juice, 2 tablespoons sugar and 1 teaspoon lemon juice to a boil in a small saucepan over medium-high heat. Cook and stir until reduced to about 2 tablespoons. Use as directed above.

Teriyaki Pork Dinner

—— Makes 4 servings ——

1 pork tenderloin (about
  16 ounces)
2 tablespoons vegetable oil,
  divided
1 large onion, thinly sliced
4 cups frozen Japanese-style
  stir-fry vegetables

1 can (8 ounces) pineapple
  chunks packed in juice
¼ cup hoisin sauce
2 tablespoons cider vinegar

1. Cut tenderloin into individual 1-inch-thick slices. Heat 1 tablespoon oil in large nonstick skillet over medium-high heat. Brown tenderloins on both sides; remove and keep warm.

2. Heat remaining 1 tablespoon oil in skillet. Add onion; cook 3 minutes or until translucent. Add vegetables; cook and stir 4 minutes.

3. Add pork with accumulated juices, pineapple with juice, hoisin sauce and vinegar to vegetable mixture; mix well. Cover; cook 5 to 7 minutes or until pork is cooked through.

Marinated Citrus Shrimp

—— Makes 16 servings ——

1 pound (about 32) large cooked shrimp, peeled and deveined (with tails on)

2 oranges, peeled and cut into segments

1 can (5½ ounces) pineapple chunks in juice, drained and ¼ cup juice reserved

2 green onions, sliced

½ cup orange juice

2 tablespoons lime juice

2 tablespoons white wine vinegar

2 tablespoons minced fresh cilantro

1 tablespoon olive or vegetable oil

1 clove garlic, minced

½ teaspoon dried basil

½ teaspoon dried tarragon

White pepper (optional)

1. Combine shrimp, orange segments, pineapple chunks and green onions in large resealable food storage bag. Mix orange juice, reserved pineapple juice, lime juice, vinegar, cilantro, oil, garlic, basil and tarragon in medium bowl; pour over shrimp mixture, turning to coat. Season with white pepper, if desired.

2. Marinate in refrigerator 2 hours or up to 8 hours.

South Pacific Pork Ribs

—— Makes 4 to 6 servings ——

2 tablespoons canola oil, divided

3½ to 4 pounds pork loin riblets (about 20 riblets)

Salt and black pepper

1 onion, chopped

1 can (20 ounces) pineapple chunks packed in juice

¼ cup all-purpose flour

½ cup water

¼ cup vinegar

¼ cup packed brown sugar

¼ cup ketchup

1 tablespoon soy sauce

Slow Cooker Directions

1. Heat 1 tablespoon oil in large skillet over medium-high heat. Season riblets with salt and pepper. Cook riblets in batches, turning to brown all sides and adding additional oil as needed. Transfer to 3½-quart slow cooker. Add onion to skillet; cook and stir 3 to 5 minutes or until softened.

2. Drain pineapple, reserving 1 cup juice. Whisk pineapple juice and flour in small bowl until well blended. Add water, vinegar, brown sugar, ketchup and soy sauce to skillet; cook and stir until blended. Stir in juice mixture until well blended; cook over medium-low heat until thickened. Stir in pineapple chunks. Pour sauce over ribs in slow cooker.

3. Cover; cook on LOW 8 to 10 hours or on HIGH 5 to 6 hours or until tender.

Mild Curry Chicken Salad with Fruit

—————— Makes 4 servings ——————

2 cups (1 pint) prepared creamy chicken salad

2 teaspoons sugar

1½ to 2 teaspoons curry powder

⅛ teaspoon ground red pepper (optional)

8 slices fresh pineapple, cut into wedges

1. Combine chicken salad, sugar, curry powder and ground red pepper, if desired, in medium bowl. Stir gently until well blended.

2. Spoon salad evenly onto plates. Arrange pineapple around salad.

Note: The curry flavor in this dish is mild, so adjust it for your taste before serving.

Variation: Add 2 tablespoons *each* currants, chopped apples, sliced red grapes, sliced green onions, and/or toasted, slivered almonds. Serve on a bed of spring greens or baby spinach leaves.

Tip: Some find the taste of curry harsh and prefer it cooked or toasted. To toast, heat a small nonstick skillet over medium-high heat. Add the curry powder and stir or tilt pan constantly to prevent burning. Cook just until fragrant, about 1 minute. Immediately remove from skillet.

Sweet and Sour Pork

——— Makes 4 servings ———

1 **pound boneless pork loin**	1 **tablespoon cornstarch**
1 **medium onion**	¼ **cup distilled white vinegar**
1 **medium green bell pepper**	3 **tablespoons light brown sugar**
2 **medium ripe peaches**	1 **tablespoon soy sauce**
1 **can (8 ounces) pineapple chunks packed in juice, undrained**	2 **tablespoons vegetable oil** **Hot cooked rice**

1. Trim fat from pork; discard. Cut pork into 1-inch cubes. Cut onion into 8 wedges. Cut bell pepper lengthwise in half. Remove stem and seeds. Rinse, dry and cut into 1-inch pieces. Cut peaches in half; discard pits. Cut each peach half into 6 wedges. Set aside.

2. Drain pineapple, reserving juice. Combine cornstarch, vinegar, brown sugar and soy sauce in small bowl. Stir in pineapple juice until smooth. Set aside.

3. Heat wok over medium-high heat 1 minute or until hot. Drizzle oil into wok and heat 30 seconds. Add pork; stir-fry about 7 minutes or until well browned. Add onion; stir-fry 2 minutes or until onion is tender. Add bell pepper; stir-fry 1 minute. Reduce heat to medium.

4. Stir juice mixture until smooth; add to wok. Add peaches and pineapple; cook and stir until sauce boils and thickens. Transfer to serving dish. Serve with rice.

Grilled Salmon
with Pineapple Salsa

—— Makes 4 servings ——

½ pineapple, cut into ½-inch cubes (about 2 cups)

½ cup Mexican beer

1 tablespoon sugar

¼ cup finely chopped red onion

¼ cup finely chopped red bell pepper

2 tablespoons chopped fresh cilantro

1 tablespoon lime juice

1 teaspoon salt, divided

4 salmon fillets (6 to 8 ounces each)

1 tablespoon olive oil

¼ teaspoon black pepper

1. Combine pineapple, beer and sugar in medium bowl; refrigerate 1 hour. Drain and discard all but 2 tablespoons liquid. Add onion, bell pepper, cilantro, lime juice and ½ teaspoon salt to pineapple mixture; refrigerate 1 hour or overnight.

2. Prepare grill for direct cooking over medium-high heat. Lightly oil grid. Rub salmon fillets with oil; sprinkle with remaining ½ teaspoon salt and black pepper.

3. Grill 5 minutes per side or until salmon just begins to flake when tested with fork. Serve salmon with salsa.

Asian-Inspired Pork & Nectarine Kabobs
—————— Makes 4 servings ——————

1 pork tenderloin (about
  1 pound)

¾ cup pineapple juice

3 tablespoons reduced-sodium
  soy sauce

1 tablespoon grated fresh ginger

1 teaspoon minced garlic

1 teaspoon ground cumin

1 teaspoon chili powder

½ teaspoon black pepper

3 fresh medium nectarines

1. Cut pork tenderloin in half lengthwise. Cut each half into 8 pieces (16 pieces total). Place pork in large resealable food storage bag. Place bag in shallow dish.

2. Stir together pineapple juice, soy sauce, ginger, garlic, cumin, chili powder and pepper. Pour over pork; seal bag. Marinate in refrigerator 3 to 6 hours.

3. Prepare grill for direct cooking.

4. Cut each nectarine into 8 chunks. Drain pork, discarding marinade. Thread pork and nectarine pieces onto 8 short skewers. Grill over medium heat 9 to 12 minutes or until pork is barely pink in center, turning once.

Sweet & Tasty Hawaiian Sandwich

— Makes 4 to 6 servings —

½ cup pineapple preserves

1 tablespoon Dijon mustard

1 round loaf (16 ounces) Hawaiian bread

8 ounces brick cheese, thinly sliced

8 ounces thinly sliced deli ham

Olive oil

Pimiento stuffed green olives (optional)

1. Combine preserves and mustard in small bowl. Cut bread in half horizontally. Pull out and discard center from bread top, leaving 1-inch shell. Spread preserves mixture on bottom half of bread. Layer with cheese, ham and top half of bread. Brush outsides of sandwich lightly with oil.

2. Heat large nonstick skillet over medium heat. Add sandwich; press down lightly with spatula. Cook sandwich 4 to 5 minutes per side or until cheese melts and sandwich is golden brown. Cut into wedges; garnish with olives.

Grilled Caramelized Salmon and Asparagus

— Makes 4 servings —

1 salmon fillet with skin (about 1 pound and 1 inch thick)

2 tablespoons packed brown sugar

1 tablespoon grated orange peel

1 teaspoon minced garlic

½ teaspoon salt

⅛ to ¼ teaspoon ground red pepper

16 asparagus spears, trimmed

¼ teaspoon black pepper

1 cup finely chopped fresh pineapple

1. Place salmon, skin side down, in shallow dish. Combine brown sugar, orange peel, garlic, salt and ground red pepper in small bowl. Rub onto salmon. Cover and refrigerate 2 to 8 hours.

2. Spray grill grid with nonstick cooking spray. Prepare grill for direct cooking.

3. Spray asparagus with cooking spray. Sprinkle with black pepper.

4. Grill salmon, skin side down, covered, over medium heat 6 minutes. Place asparagus on grid. Grill, covered, turning asparagus occasionally, 4 minutes or until salmon begins to flake when tested with fork and asparagus begins to brown.

5. Cut salmon into four equal pieces. Top salmon with pineapple and serve with asparagus.

Shrimp and Pineapple Kabobs

Makes 4 servings

½ pound medium raw shrimp, peeled and deveined (with tails on)

½ cup pineapple juice

¼ teaspoon garlic powder

12 chunks canned pineapple

1 green bell pepper, cut into 1-inch pieces

¼ cup prepared chili sauce

1. Combine shrimp, juice and garlic powder in medium bowl; toss to coat. Marinate in refrigerator 30 minutes. Drain shrimp; discard marinade.

2. Alternately thread pineapple, bell pepper and shrimp onto 4 (10-inch) skewers. Brush with chili sauce. Grill 4 inches from heat 5 minutes or until shrimp are opaque, turning once and basting with chili sauce.

Pineapple Basil Chicken Supreme

Makes 4 servings

1 can (8 ounces) pineapple chunks in unsweetened juice

2 teaspoons cornstarch

2 tablespoons peanut oil

1 pound boneless skinless chicken breasts, cut into ¾-inch pieces

2 to 4 serrano peppers,* cut into thin strips (optional)

2 cloves garlic, minced

2 green onions, cut into 1-inch pieces

¾ cup roasted unsalted cashew nuts

¼ cup chopped fresh basil (do not use dried)

1 tablespoon fish sauce**

1 tablespoon soy sauce

   Hot cooked rice (optional)

*Serrano peppers can sting and irritate the skin, so wear rubber gloves when handling peppers and do not touch your eyes.

**Fish sauce is available at most large supermarkets and Asian markets.

1. Drain pineapple, reserving juice. Combine reserved juice and cornstarch in small bowl; set aside.

2. Heat wok or large skillet over high heat 1 minute. Drizzle oil into wok; heat 30 seconds. Add chicken, serrano peppers, if desired, and garlic; stir-fry 3 minutes or until chicken is cooked through. Add green onions; stir-fry 1 minute.

3. Stir cornstarch mixture; add to wok. Cook and stir 1 minute or until thickened. Add pineapple, cashews, basil, fish sauce and soy sauce; cook and stir 1 minute or until heated through. Serve over rice, if desired.

SIPPERS & SMOOTHIES

Ginger-Pineapple Spritzer

—— Makes 3 cups or 4 servings ——

2 cups pineapple juice or cranberry juice

1 tablespoon chopped crystallized ginger

1 cup chilled club soda or sparkling water

Ice cubes

Fresh pineapple wedges and mandarin orange segments (optional)

1. Combine pineapple juice and ginger in small saucepan. Bring to a simmer. Pour into small pitcher. Cover; refrigerate for 8 to 24 hours.

2. Strain juice mixture; discard ginger. Gently stir club soda into juice mixture. Serve in glasses over ice cubes. Garnish with pineapple wedges and mandarin orange segments.

Tropical Breakfast Smoothie

—————— Makes 4 servings ——————

1 can (20 ounces) pineapple chunks in juice, undrained

1 ripe banana

½ cup ice cubes

½ cup orange juice

¼ cup flaked coconut

1 tablespoon fresh lime juice

Maraschino cherries and lime wedges (optional)

1. Combine pineapple, banana, ice, orange juice, coconut and lime juice in blender; blend until smooth.

2. Pour into 4 glasses. Garnish with cherries and lime wedges.

Tip: If all of the ingredients will not fit in the blender, purée all ingredients except the banana. Add banana and purée again.

Guava Fruit Punch

—— Makes 4 servings ——

1½ cups boiling water
2 decaffeinated tea bags
3 thin slices peeled fresh ginger
2 cups guava juice

¾ cup pineapple juice
1 to 2 tablespoons lemon juice
Ice cubes

1. Combine boiling water, tea bags and ginger in heatproof pitcher; steep 5 minutes. Discard tea bags and ginger. Cool to room temperature.

2. Add guava juice, pineapple juice and lemon juice to tea mixture; mix well. Serve in tall glasses over ice.

Caribbean Dream

—— Makes 2 servings ——

¾ cup low-fat vanilla ice cream

¾ cup pineapple sherbet

¾ cup tropical fruit salad, drained

¼ cup frozen banana-orange juice concentrate

¼ teaspoon rum extract

Fresh pineapple wedges

1. Combine ice cream, sherbet, fruit salad, juice concentrate and extract in blender. Process 1 to 2 minutes or until smooth.

2. Pour into 2 glasses. Serve immediately. Garnish with fresh pineapple wedges.

Serve It With Style!: Try adding a tablespoon of rum instead of rum-flavored extract for a more mature flavor.

Island Delight Smoothie

—————— Makes 4 servings ——————

2 cups chopped fresh or jarred mango

1 container (16 ounces) plain nonfat yogurt

1½ cups pineapple-orange juice, chilled

1 cup chopped pineapple

1 frozen banana

½ cup sliced fresh strawberries

2 tablespoons honey

1½ cups ice cubes

Fresh banana slices (optional)

1. Combine mango, yogurt, pineapple-orange juice, pineapple, frozen banana, strawberries, honey and ice in blender; process until fruit is puréed and mixture is smooth.

2. Pour into 4 glasses. Garnish with banana slices. Serve immediately.

SWEETS & TREATS

Cream Cheese Cupcakes
—————— Makes 24 cupcakes ——————

3 packages (8 ounces each) cream cheese, softened

5 eggs

1¼ cups sugar, divided

2½ teaspoons vanilla, divided

1 container (16 ounces) sour cream

1 cup chopped fresh pitted cherries, fresh blueberries and/or canned crushed pineapple, drained

1. Preheat oven to 325°F. Line 24 standard (2½-inch) muffin cups with paper baking cups.

2. Beat cream cheese, eggs, 1 cup sugar and 1½ teaspoons vanilla in large bowl with electric mixer at medium speed 2 minutes or until well blended. Fill prepared muffin cups three fourths full.

3. Bake 20 minutes or until light golden brown. Cool cupcakes in pans 5 minutes. (Centers of cupcakes will sink slightly.) Do not remove cupcakes from pans.

4. Meanwhile, combine sour cream, remaining ¼ cup sugar and 1 teaspoon vanilla in medium bowl; stir until blended. Fill depression in cupcakes with sour cream mixture. Bake 5 minutes. Cool cupcakes in pans 10 minutes; remove to wire racks to cool completely.

5. Top cupcakes with fruit as desired.

Tropical Dump Cake

— Makes 12 to 16 servings —

1 can (20 ounces) crushed pineapple, undrained

1 can (15 ounces) peach slices in light syrup, undrained

1 package (about 15 ounces) yellow cake mix

½ cup (1 stick) butter, cut into thin slices

1 cup packed brown sugar

½ cup flaked coconut

½ cup chopped pecans

1. Preheat oven to 350°F. Spray 13×9-inch baking pan with nonstick cooking spray.

2. Spread pineapple and peaches in prepared pan. Top with cake mix, spreading evenly. Top with butter in single layer, covering cake mix as much as possible. Sprinkle with brown sugar, coconut and pecans.

3. Bake 40 to 45 minutes or until toothpick inserted into center of cake comes out clean. Cool at least 15 minutes before serving.

Marinated Pineapple Dessert

—————— Makes 4 servings ——————

1 can (20 ounces) pineapple chunks in juice, undrained

2 tablespoons honey

1 stick cinnamon, broken into pieces

1 tablespoon lemon juice

1 teaspoon vanilla

½ teaspoon grated lemon peel

1½ cups strawberries, cut into halves

¼ cup flaked coconut, toasted*

To toast coconut, spread in single layer in heavy-bottomed skillet. Cook over medium heat 1 to 2 minutes until lightly browned, stirring frequently. Remove from skillet immediately. Cool before using.

1. Combine pineapple with juice, honey, cinnamon, lemon juice, vanilla and lemon peel in small saucepan. Bring to a boil over medium-high heat. Pour mixture into medium bowl; cover and refrigerate at least 4 hours or up to 24 hours.

2. Drain pineapple mixture; reserve liquid. Remove and discard cinnamon pieces.

3. Divide pineapple and strawberries evenly among 4 dessert dishes. Pour reserved liquid evenly over fruit. Sprinkle with coconut before serving.

Pineapple Cheesecake
—— Makes 10 to 12 servings ——

1 package (3 ounces) lemon-flavored gelatin

1½ cups hot water

1 package (8 ounces) cream cheese, softened

1 cup sugar

½ cup milk

1 teaspoon vanilla

1 can (20 ounces) crushed pineapple, drained

1 container (8 ounces) frozen whipped topping, thawed

3 cups graham cracker crumbs

4 tablespoons powdered sugar

⅓ cup butter, melted

1. Dissolve gelatin in hot water in small bowl. Set aside to cool.

2. Beat cream cheese and sugar in large bowl with electric mixer at medium speed until creamy. Add milk and vanilla; beat until well combined. Add pineapple and gelatin mixture to cream cheese mixture. Fold in whipped topping. Set aside.

3. Combine cracker crumbs and powdered sugar in medium bowl. Add melted butter and mix until crumbly. Reserve 1¼ cups graham cracker mixture for topping. Spread remaining graham cracker mixture in 11×7-inch pan. Pat down firmly to form bottom crust.

4. Pour cream cheese mixture into pan. Sprinkle remaining 1¼ cups graham cracker mixture over top of cream cheese mixture.

5. Refrigerate 4 hours or until firmly set.

Serving Suggestion: Garnish with pineapple slices or maraschino cherries. Or, top with 1 can (21 ounces) cherry pie filling.

Piña Colada Whoopie Pies

—————— Makes 20 whoopie pies ——————

Cookies

- 2 cups all-purpose flour
- 1 teaspoon baking powder
- ¾ teaspoon baking soda
- ¼ teaspoon salt
- ½ cup (1 stick) butter, softened
- ¼ cup packed brown sugar
- 2 eggs
- 2 teaspoons rum extract
- ¼ cup cream of coconut
- 1 can (about 8 ounces) crushed pineapple, undrained
- 1½ cups sweetened flaked coconut

Filling

- 3½ cups powdered sugar
- 1 package (8 ounces) cream cheese, softened
- 2 teaspoons rum extract
- ¼ cup sweetened flaked coconut

1. For cookies, preheat oven to 350°F. Line 2 cookie sheets with parchment paper. Combine flour, baking powder, baking soda and salt in medium bowl.

2. Beat butter and brown sugar in large bowl with electric mixer at medium speed about 2 minutes or until creamy. Add eggs and 2 teaspoons rum extract; beat 1 minute or until well blended.

3. Add half of flour mixture; beat just until blended. Beat in cream of coconut. Add remaining flour mixture and pineapple; beat 2 minutes or until well blended. Fold in 1½ cups coconut. Drop 2 level tablespoonfuls of batter 2 inches apart onto prepared cookie sheets.

4. Bake 11 to 13 minutes or until tops spring back when lightly touched. Cool 10 minutes on cookie sheets. Remove to wire racks; cool completely.

5. For filling, beat powdered sugar and cream cheese in large bowl with electric mixer at medium speed about 2 minutes or until light and fluffy. Add 2 teaspoons rum extract; beat 1 minute.

6. Pipe or spread 2 rounded tablespoons filling on flat side of half of cookies; top with remaining cookies. Place ¼ cup coconut in shallow dish; roll edges of cookies in coconut to coat lightly.

Caramelized Pineapple

———— Makes 4 servings ————

1 tablespoon margarine

2 cups fresh pineapple chunks

3 tablespoons sugar

¾ cup vanilla reduced-fat frozen yogurt

1. Spray baking sheet with nonstick cooking spray.

2. Melt margarine in large nonstick skillet over medium-high heat. Add pineapple and sugar; cook and stir about 7 minutes or until pineapple begins to brown. Cook 2 to 4 minutes more or until golden brown, stirring and turning pineapple occasionally. Spread on prepared baking sheet. Cool 5 minutes.

3. Spoon pineapple into 4 dessert dishes. Top each with small scoop of frozen yogurt. Serve immediately.

Easy Upside Down Cake

Prep Time: 15 minutes Bake Time: 40 minutes

——— Makes 10 servings ———

1 can (20 oz.) DOLE® Pineapple Slices

¼ cup butter or margarine, melted

⅔ cup packed brown sugar

10 maraschino cherries

1 package (2 layer) yellow or pineapple-flavored cake mix

- **Preheat** oven to 350°F.

- **Drain** pineapple, reserving ¾ cup juice.

- **Stir** together melted butter and brown sugar in 12-inch skillet with heatproof handle. Arrange pineapple slices in sugar mixture. Place cherry in center of each pineapple slice.

- **Prepare** cake mix according to package directions, replacing water with reserved ¾ cup juice. Pour batter evenly over pineapple.

- **Bake** 40 to 45 minutes or until toothpick inserted in center comes out clean.

- **Cool** 5 minutes. Loosen edges and invert onto serving platter.

Note: Cake can be baked in 13×9-inch baking pan instead of skillet. Prepare and assemble cake as above except cut two pineapple slices in half and place whole slices along edges of pan and halved slices in center. Place cherries in center of slices. Bake and cool as above.

Mini Upside Down Cakes: Drain 1 can (20 oz.) DOLE® Crushed Pineapple, reserving juice. Grease 24 muffin cups. Stir ⅓ cup melted butter with ⅔ cup packed brown sugar. Evenly spoon mixture into bottoms of cups; spoon about 1 tablespoon crushed pineapple over sugar mixture. Prepare cake mix as above. Evenly pour batter into cups. Bake 20 to 25 minutes. Invert onto serving platter. Makes 24 servings.

INDEX

Acknowledgment

The publisher would like to thank the following company for the use of their recipe in this publication.

Dole® Food Company

METRIC CONVERSION CHART

VOLUME MEASUREMENTS (dry)

$\frac{1}{8}$ teaspoon = 0.5 mL
$\frac{1}{4}$ teaspoon = 1 mL
$\frac{1}{2}$ teaspoon = 2 mL
$\frac{3}{4}$ teaspoon = 4 mL
1 teaspoon = 5 mL
1 tablespoon = 15 mL
2 tablespoons = 30 mL
$\frac{1}{4}$ cup = 60 mL
$\frac{1}{3}$ cup = 75 mL
$\frac{1}{2}$ cup = 125 mL
$\frac{2}{3}$ cup = 150 mL
$\frac{3}{4}$ cup = 175 mL
1 cup = 250 mL
2 cups = 1 pint = 500 mL
3 cups = 750 mL
4 cups = 1 quart = 1 L

VOLUME MEASUREMENTS (fluid)

1 fluid ounce (2 tablespoons) = 30 mL
4 fluid ounces ($\frac{1}{2}$ cup) = 125 mL
8 fluid ounces (1 cup) = 250 mL
12 fluid ounces (1$\frac{1}{2}$ cups) = 375 mL
16 fluid ounces (2 cups) = 500 mL

WEIGHTS (mass)

$\frac{1}{2}$ ounce = 15 g
1 ounce = 30 g
3 ounces = 90 g
4 ounces = 120 g
8 ounces = 225 g
10 ounces = 285 g
12 ounces = 360 g
16 ounces = 1 pound = 450 g

DIMENSIONS

$\frac{1}{16}$ inch = 2 mm
$\frac{1}{8}$ inch = 3 mm
$\frac{1}{4}$ inch = 6 mm
$\frac{1}{2}$ inch = 1.5 cm
$\frac{3}{4}$ inch = 2 cm
1 inch = 2.5 cm

OVEN TEMPERATURES

250°F = 120°C
275°F = 140°C
300°F = 150°C
325°F = 160°C
350°F = 180°C
375°F = 190°C
400°F = 200°C
425°F = 220°C
450°F = 230°C

BAKING PAN SIZES

Utensil	Size in Inches/Quarts	Metric Volume	Size in Centimeters
Baking or Cake Pan (square or rectangular)	8×8×2	2 L	20×20×5
	9×9×2	2.5 L	23×23×5
	12×8×2	3 L	30×20×5
	13×9×2	3.5 L	33×23×5
Loaf Pan	8×4×3	1.5 L	20×10×7
	9×5×3	2 L	23×13×7
Round Layer Cake Pan	8×1½	1.2 L	20×4
	9×1½	1.5 L	23×4
Pie Plate	8×1¼	750 mL	20×3
	9×1¼	1 L	23×3
Baking Dish or Casserole	1 quart	1 L	—
	1½ quart	1.5 L	—
	2 quart	2 L	—